BR STANDARD
FREIGHT WAGONS
A PICTORIAL SURVEY

DAVID LARKIN

D. BRADFORD BARTON LIMITED

introduction

At the time of Nationalisation on 1 January 1948, British Railways inherited a vast number of wagons from the existing companies and from private owners. On paper, there was no need for the introduction of standard types to replace this acquired stock. However, two important points motivated BR when it was decided to scrap much of the latter and replace them with new designs.

Firstly, the economic climate of the immediate post-war period drew a false picture of traffic trends. Petrol rationing masked the fact that freight traffic was still going to be won by road hauliers, as it had been pre-war. Thus BR felt able to continue the development of various innovations, such as shock-absorbing equipment which, but for the war, would have been introduced generally by the companies.

Secondly, much of the inherited stock was old-fashioned and inefficient. The use of vacuum-brakes was not common and some of the wagons with this form of brake were unsuitable for fast trains. Much mineral traffic was still being conveyed in damage-prone wooden-bodied stock and this made the introduction of large numbers of steel-bodied wagons essential.

For the first two years of its existence, BR built batches of wagons that were almost identical to pre-Nationalisation designs. Many were probably ordered by the companies anyway. Then, in 1951, the first of the standard designs began to appear. They did not represent any great advance in wagon design, being merely improved versions of what had gone before. The real revolution came in the late 1960s when it was realised that the railways were fighting a losing battle to retain what freight traffic they had not already given away. This period is beyond the scope of this present volume and the types illustrated here are those built by BR from 1948 to 1968.

In addition to standardising wagon designs, liveries and number series were also rationalised. The inherited stock was dealt with as follows—

Company stock: original number retained and prefixed by a single letter appertaining to company of origin (i.e., E – (Ex. LNER); M – (Ex. LMS); S – (Ex. SR); W – (Ex. GWR).

Privately owned stock: new number selected at random from 1–399999 approx. and prefixed P.

BR-built stock: number selected from block allocated to class of wagon and prefixed B. (N.B.: This applied to general traffic stock).

BR-built Engineers Dept. stock: all stock had numbers prefixed DB. Stock originally general traffic but transferred to Engineers Dept. sometimes had DB painted on the wagon but retained original number plate with B prefix.

The following standard liveries were applied to all freight rolling stock used by British Railways—

Non-fitted stock: black or grey roof (where applicable), grey bodywork, black solebar (and below), white lettering on black panels.

Vacuum-fitted stock: black or grey roof (where applicable), bauxite (red-brown) bodywork, black solebar (and below), white lettering.

These liveries are now seen on all vehicles except as mentioned below. There is considerable variation in shade of paint used, especially with non-fitted stock. Some of the latter is painted in vacuum-fitted livery, and vice versa. Many steel-bodied vehicles, having suffered knocks to the paintwork, present a brown appearance due to rust. The special livery (excluding Engineers Dept.) applicable to insulated stock was black or grey roof, white bodywork, black solebar (and below), black lettering. This was changed in 1957 to black or grey roof, ice-blue (pale-blue) bodywork, black solebar (and below), white lettering.

The lettering style used by BR was somewhat similar to the system adopted during the 1930s by all companies, the number being positioned on the left, just above the solebar, with the tare-weight to the right. Code-name, if any, was generally above the number. A later style, introduced about 1957, grouped all the lettering to the left with surrounding white lines.

Although there were a number of design failures and types whose original function rapidly became outmoded, the majority of standard designs built by BR can be considered successful. Nevertheless, they did not represent positive progress in wagon design and consequently, when road competition forced BR to introduce new policies, such as the Freightliner system, they gradually became redundant. The current situation on British Rail, with regard to the wagon types illustrated in this volume, is one of a gradual run-down. Scrapping of general-purpose wagons has been widespread as new types have come in. In the specialised traffic field, the original standard designs, in company with pre-Nationalisation stock, are still used but new types are being introduced, and they will not last much longer. Thus this volume presents a record of modern stock which will shortly vanish from the railway scene.

In choosing particular examples for the illustrations, the intention has been to select those which will be of particular interest not only to the railway enthusiast but also to the modeller, not least in that they show freight rolling stock in day-to-day service condition. For modelling purposes this is of more use, when realism is sought, than the usual pristine ex-works livery which bears little resemblance to the appearance of wagons after months or years of use.

Further volumes in preparation in this series will cover pre-Nationalisation freight stock used on BR, private-owner wagons, and general BR parcels stock.

© copyright D. Bradford Barton Ltd 1979.   ISBN 0 85153 240 3

printed in Great Britain by H. E. Warne Ltd, London and St. Austell

for the publishers

D. BRADFORD BARTON LTD · Trethellan House · Truro · Cornwall · England

contents

	page
16-ton non-fitted mineral wagon (welded)	5
16-ton non-fitted mineral wagon (rivetted)	5
16-ton vacuum-fitted mineral wagon	6
21-ton non-fitted mineral wagon (ex-private owner)	6
21-ton non-fitted mineral wagon (standard design)	7
21-ton vacuum-fitted mineral wagon (standard design)	7
24½-ton non-fitted mineral wagon (standard design)	8
24½-ton non-fitted mineral wagon (non-standard)	8
21-ton non-fitted coal hopper (LNER design)	9
21-ton non-fitted coal hopper wagon (standard design)	9
21-ton vacuum-fitted coal hopper wagon (standard design)	10
21-ton non-fitted hopper wagon (experimental version)	10
24½-ton coal hopper wagon (standard design; non-fitted)	11
32-ton air-braked coal hopper wagon (HOP32AB design)	11
20-ton coke hopper wagon (LMS design)	12
20-ton coke hopper wagon (BR design)	12
24-ton non-fitted iron ore hopper wagon	13
13-ton non-fitted iron ore hopper wagon	13
25½-ton non-fitted sand hopper wagon	14
30-ton vacuum-fitted hopper wagon	14
33-ton vacuum-fitted iron ore hopper wagon	15
25-ton vacuum-fitted anhydrite hopper wagon	15
26-ton iron ore tippler wagon	16
13-ton sand tippler wagon	16
13-ton china clay open wagon (standard design)	17
11-ton Conflat L with L-type containers	17
22-ton pressure-discharge bulk powder wagon (Presflo)	18
20-ton pressure-discharge bulk powder wagon (Prestwin)	18
24-ton bulk powder wagon (Covhop: standard design)	19
13-ton bulk powder open wagon (special type)	19
21-ton Conflat LD wagon with LD-type containers	20
13-ton Conflat A with LF-type container	20
20-ton bulk grain covered hopper wagon	21
21-ton bulk grain open hopper wagon (special type)	21
13-ton open wagon (medium-height; dropside)	22
13-ton open wagon (high-sided; all wooden bodywork)	22
13-ton open wagon (high-sided; with steel ends)	23
13-ton open wagon (high-sided; LNER design)	23
12-ton shock-absorbing open wagon (high-sided; wooden-bodied)	24
12-ton shock-absorbing open wagon (high-sided; with steel ends)	24
12-ton shock-absorbing open wagon (standard design; late variant)	25
13-ton shock-absorbing open wagon	25
12-ton ventilated van (vacuum-fitted; GWR plywood design)	26
12-ton ventilated van (vacuum-fitted; SR plywood design)	26
12-ton ventilated van (vacuum-fitted; LMS planked design)	27
12-ton ventilated van (vacuum-fitted; LMS plywood design)	27
12-ton ventilated van (vacuum-fitted; standard design, first pattern)	28
12-ton ventilated van (vacuum-fitted; standard design, fourth pattern)	28
12-ton ventilated van (vacuum-fitted; wide-opening doors)	29
12-ton shock-absorbing ventilated van (GWR design)	29
12-ton shock-absorbing ventilated van (LNER design)	30
12-ton shock-absorbing ventilated van (LMS design)	30
12-ton shock-absorbing ventilated van (standard design; hybrid)	31
12-ton shock-absorbing ventilated van (standard design; third pattern)	31
12-ton ventilated pallet van (offset doors)	32
12-ton ventilated pallet van (sliding doors)	32
12-ton shock-absorbing ventilated pallet van	33
24-ton long-wheelbase air-braked pallet van	33
11-ton Conflat A, with A-type furniture container	34

	page		page
13-ton Conflat A, with B-type furniture container	34	13-ton single bolster wagon (non-fitted; standard design)	50
13-ton Conflat A, with BK-Type furniture container	35	22-ton double bolster wagon	51
Modified 13-ton Conflat A, modified BD-type container	35	26-ton twin-bolster wagon set (Lowfit conversion)	51
Modified 13-ton Conflat A, with Speedfreight container	36	30-ton bogie bolster wagon (BBC; non-fitted)	52
12-ton vacuum-fitted cattle wagon (SR design; modified)	36	30-ton bogie bolster wagon (BBC; vacuum-fitted)	52
8-ton vacuum-fitted cattle wagon (standard design)	37	42-ton bogie bolster wagon (BBD; non-fitted)	53
10-ton vacuum-fitted ventilated meat van	37	42-ton bogie bolster wagon (BBD; vacuum-fitted)	53
BM-type ventilated meat container	38	50-ton bogie bolster wagon (Borail)	54
10-ton vacuum-fitted insulated meat van	38	13-ton vehicle-carrying open wagon (Lowfit; LNER design)	54
FM-type insulated fish and meat container	39	13-ton vehicle-carrying open wagon (Lowfit; standard design)	55
12-ton insulated fish van (passenger; LNER pattern)	39	8-ton bogie vehicle-carrying open wagon (Bocar P)	55
12-ton insulated fish van (passenger; standard pattern)	40	23-ton vehicle-carrying well wagon (Lowmac WE; GWR design)	56
11-ton Conflat A (with AFU-type insulated container)	40	25-ton vehicle-carrying well wagon (Lowmac EU; modified LNER design)	56
12-ton ventilated fruit van (vacuum-fitted; GWR design)	41	38-ton vehicle-carrying bogie well wagon (Rectank EB)	57
12-ton ventilated fruit van (vacuum-fitted; LMS design)	41	20-ton vehicle-carrying trolley wagon (Flatrol SB)	57
12-ton ventilated fruit van (vacuum-fitted: LNER design)	42	50-ton vehicle-carrying bogie trolley wagon (Flatrol EY)	58
12-ton ventilated fruit van (vacuum-fitted; standard)	42	12-ton non-fitted glass carrying wagon	58
8-ton insulated banana van (LNER pattern)	43	16-ton vacuum-fitted pallet brick-carrying wagon	59
12-ton insulated banana van (standard design)	43	50-ton bogie trolley wagon (Flatrol MJ)	59
12-ton gunpowder van (standard design)	44	15-ton vacuum-fitted timber-carrying wagon	60
22-ton plate wagon (standard design; non-fitted)	44	22-ton vacuum-fitted ale pallet wagon	60
22-ton plate wagon (standard design; vacuum-fitted)	45	Vacuum-fitted diesel brake tender (early type)	61
42-ton bogie plate wagon (non-fitted; standard design)	45	Vacuum-fitted diesel brake tender (standard type)	61
42-ton bogie plate wagon (vacuum-fitted; standard design)	46	20-ton brake van (non-fitted; LMS design)	62
21-ton trestle plate wagon (non-fitted; 4-wheeled type)	46	20-ton brake van (vacuum-piped; LMS design)	62
50-ton bogie trestle well wagon (vacuum-fitted)	47	20-ton brake van (GWR design; non-fitted)	63
13-ton short-wheelbase pipe wagon (with drop sides)	47	20-ton brake van (standard design; non-fitted)	63
12-ton pipe wagon (LNER design)	48	20-ton brake van (standard design; vacuum-fitted)	64
22-ton tube wagon (vacuum-fitted; LNER design)	48	20-ton brake van (modified standard design)	64
22-ton tube wagon (non-fitted; GWR design)	49		
22-ton tube wagon (non-fitted; standard design)	49		
22-ton tube wagon (vacuum-fitted; standard design)	50		

16-TON NON-FITTED MINERAL WAGON [WELDED]
B96956 [Built 1954, Pressed Steel, Lot 2258] is representative of the commonest type of non-fitted 16-ton mineral wagon on BR, following LMS practice although with simplified brakegear. In common with other early vehicles in this class, it lacks drop flaps above the side doors. The painting of the solebar and buffer-beam in the same colour as the bodywork is unusual. The load of metal waste on this occasion illustrates the fact that this type is no longer exclusively used on coal traffic.
[Cargo Fleet, Middlesbrough 1969]

16-TON NON-FITTED MINERAL WAGON [RIVETTED] The rivetted version of the 16-ton mineral wagon, derived from LNER practice, is less common than the welded version and is represented here by B267453 [Built 1956, Hurst Nielson, Lot 2811]. The pattern of end door found on this vehicle is commonest on the rivetted version but is sometimes replaced by the welded pattern, as seen on B96956. One noteworthy identifying feature is the rounded top edge to the bodywork. Welded vehicles have U-shaped edges. [Radstock, Somerset 1971]

16-TON VACUUM-FITTED MINERAL WAGON Vacuum-fitted 16-ton mineral wagons first appeared in the mid-1950's and were at first rated XP, a classification which allows goods stock to run at passenger train speeds. The 9' wheelbase was found unsuitable for high-speed running however and the classification was withdrawn. B550220 [Built 1958; Birmingham RCW; Lot 2907] has the later pattern brakegear which appears on about one-third of all vacuum-fitted 16-ton mineral wagons. A number of non-fitted wagons were converted to vacuum-fitted, following the pattern shown here. Conversely some vacuum-fitted stock has appeared without pipes or cylinders. [*Hoo Junction, Strood 1970*]

21-TON NON-FITTED MINERAL WAGON [Ex-PRIVATE OWNER] In the 1930's, the GWR encouraged the use of high-capacity coal wagons in South Wales and hired out steel-bodied vehicles of 21 tons capacity to many of the collieries in that area. This stock came into the BR fleet at nationalisation and is numerous enough to be regarded as a standard type. P339298K [Built 1944, Registration No. GWR 7225] is typical of the stock. No attempt was made to re-number these wagons into a separate series, numbering being random between P1 and P399999. The only noteworthy visual difference between wagons of this class is the overall height; the vehicle to the left shows this point. [*Newbridge, Glamorgan 1972*]

21-TON NON-FITTED MINERAL WAGON [STANDARD DESIGN] BR developed the GWR idea of high-capacity coal wagons for general use and produced 21 tons capacity stock of similar basic design but with features common to the 16-ton mineral wagon class. Both welded and rivetted bodywork was used in the construction of the type and B200825 [Built 1951, Metropolitan Cammell, Lot 2190] is typical of the latter version. A number of wagons of both versions received roller-bearings. This stock can be seen in local coal-yards everywhere; the poor quality of the fuel carried by this example indicates that it was destined for industrial use. [*Stanton Gate, Notts. 1971*]

21-TON VACUUM-FITTED MINERAL WAGON [STANDARD DESIGN] This stock is a supreme example of BR standardisation policy, all vehicles produced being to the design illustrated by B312825 [Built 1962, BR Derby, Lot 3430]. Most of this class bore the legend 'TO WORK WITHIN SOUTH WALES AND MONMOUTHSHIRE ONLY', an echo of the original GWR policy regarding high-capacity coal wagons. They were originally intended for block-working supplying industry but they are now seen in all parts of the country, often in local coal-yards. [*Chichester, Sussex 1969*]

24½-TON NON-FITTED MINERAL WAGON [STANDARD DESIGN] To supplement the rather motley collection of Loco Coal wagons inherited from the Grouping companies at nationalisation, BR produced an enlarged version of the standard 21-ton non-fitted mineral wagon design. There was not, however, the variety of bodywork found in the 21 tons capacity stock, all 24½-ton mineral wagons having welded bodywork. None has received vacuum-brakegear. The first use of this stock was serving large motive power depots in block-train formation. Later they supplied power stations in a similar manner but they have now been relegated to ordinary coal traffic. B281478N [Built 1955, BR Shildon, Lot 2600] is a typical standard wagon. [*Hoo Junction, Strood 1969*]

24½-TON NON-FITTED MINERAL WAGON [NON-STANDARD VARIATION] Certain of the wagons belonging to the 24½-ton mineral wagon class have non-standard features. A fairly common variation is the use of roller-bearings. Some have side doors of a pressed pattern and odd combinations can be seen using both this pattern and the welded type found on B281478N. Rather less common is the variation illustrated by B280471N [Built 1953, BR Shildon, Lot 2460]. This involves the removal of the side doors and the drop flaps normally found above them. To prevent the despatch of one of these wagons to a destination without wagon tippler facilities, the legend 'NO DOORS' is painted in a prominent position. [*Stanton Gate, Notts. 1971*]

21-TON NON-FITTED COAL HOPPER WAGON [LNER DESIGN] Within its own territory, the North Eastern Railway encouraged the use of bottom-discharge vehicles in conjunction with coal-staithes. Wooden stock was built and, to replace these as they became time-expired, the LNER developed a high-capacity steel design. BR continued the development and built batches of wagons to the LNER design. Both welded and rivetted versions were built, B410140K [Built 1950, Cravens, Lot 2163] being one of the former; all have the tall LNER-style brake lever. [*Hoo Junction, Strood 1968*]

21-TON NON-FITTED COAL HOPPER WAGON [STANDARD DESIGN] B421818K [Built 1957; Metropolitan Cammell; Lot 2935] represents the standard BR-designed wagon with welded bodywork and low brake-lever. Roller bearings have been used on a number of wagons. Although a standard vehicle, this particular example has a special livery. In addition to the normal non-fitted wagon livery, it bears the legend 'CHARRINGTONS' in black on a red background. For a number of years, the depots of this well-known firm of coal merchants were served by block-trains of stock bearing this or a similar livery. [*Hoo Junction, Strood 1968*]

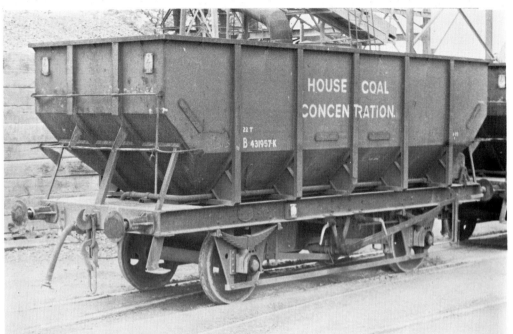

21-TON VACUUM-FITTED COAL HOPPER WAGON [STANDARD DESIGN]
Vacuum-fitted vehicles of the 21-ton Coal Hopper class were introduced during the mid-1950's and B431957 [Built 1956; Pressed Steel; Lot 3157] is a typical example of the only design built. Many bore the legend 'HOUSE COAL CONCENTRATION' when new, due to the closing of the smaller coal yards in large towns and cities and concentrating this traffic on a larger coal depot with conveyor belts and hopper facilities. During the early 1970's many have had the lettering painted out and have been used for carrying sand. This example, in use for this, has not yet had its livery modified. [*Ardingly, Sussex 1970*]

21-TON NON-FITTED HOPPER WAGON [EXPERIMENTAL VERSION]
B422416K [Built 1958; Fairfield S & E; Lot 3013] is an experimental vehicle modified from the standard 21-ton Coal Hopper design by the fitting of an overall roof with sliding panels at each end. These panels are operated by the white-painted hand-wheels at each end of the solebar. This vehicle bears no indication of the load it is intended to carry but would seem to duplicate to some extent the Covhop type. [*Hoo Junction, Strood 1968*]

24½-TON COAL HOPPER WAGON [STANDARD DESIGN; NON-FITTED]

Development of the 21-ton mineral and 21-ton Coal Hopper designs continued on similar lines and, in common with the 24½-ton mineral type, an enlarged coal hopper, with a capacity for 24½ tons, was also introduced. B338075N [Built 1965; Pressed Steel; Lot 3525] represents the non-fitted examples of this stock. Some early vehicles have oil-axleboxes but the majority have roller-bearings whilst a number are vacuum-fitted. This stock replaced the 24½-ton mineral on power station workings and have themselves now been superceded by the HOP 32 AB type. This particular wagon has a livery oddity in that the lettering is painted on a bauxite panel. [*Stanton Gate, Notts. 1971*]

32-TON AIR-BRAKED COAL HOPPER WAGON [HOP32AB DESIGN]

In the 1960's a radical re-think of BR wagon design occurred and new types began to appear, the two major features common to all these being air-brakes and a long wheelbase. Most of these new types are beyond the scope of this present volume because they represent a new era of standard wagon design and not all types have yet appeared. The HOP32AB, however, is the link between the old and new eras and retains the prefix letter to the wagon number, the only air-braked type to do so. B351678 [Built 1966; BR Shildon; Lot 3528] is uncommon in being fitted with extensions to the sides and ends. [*Aylesbury, Bucks. 1970*]

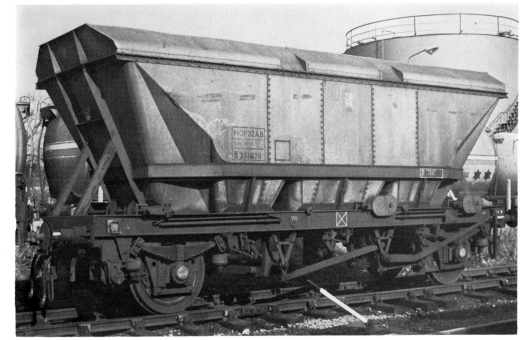

20-TON COKE HOPPER WAGON [LMS DESIGN] The LMS designed this class of vehicle and built small numbers. Some were also built for private owners and subsequently came to BR. The first BR-built stock closely followed the original design and, despite the lettering found on B447246 [Built 1950; BR Shildon; Lot 2039], had the same 20-ton capacity. This class is mainly for factory and foundry traffic but might occasionally be seen in local coal yards. [*Rogerstone, Glamorgan 1972*]

20-TON COKE HOPPER WAGON [BR DESIGN] Subsequent BR production of 20-ton Coke Hopper wagons followed two basic variations on the original LMS design, although in both these the shape and dimensions of the vehicle remained the same. At first, vehicles which had solid ends and wooden side raves were produced in non-fitted and vacuum-fitted versions. Later vehicles had solid bodywork, as illustrated by B449096 [Built 1958; BR Shildon; Lot 3122]. This particular wagon, being non-fitted, should be grey but is extremely rusty. Some of this variety have the legend 'PENSNETT'. Many of the later vehicles, of both versions, have roller-bearings. [*Toton, Notts. 1971*]

24-TON NON-FITTED IRON ORE HOPPER WAGON [STANDARD DESIGN]

Vehicles of this type were introduced by Charles Roberts of Wakefield in the early 1930's and BR supplemented those it inherited from various private owners with further batches to the same design. B435475 [Built 1949, Charles Roberts; Lot 2055] is typical of the BR-built batches. A small number, both from inherited stock and BR-built stock, have been fitted with roller-bearings. In addition to iron ore, this type also carries sand. [*Ipswich, Suffolk 1969*]

13-TON NON-FITTED IRON ORE HOPPER WAGON [STANDARD DESIGN]

As referred to under '21-ton Coal Hopper Wagons (LNER design)', the North Eastern Railway built wooden hopper wagons for coal traffic. In a similar manner, smaller vehicles were constructed for iron ore and the type represented here by B400914 [Built 1950; BR Shildon; Lot 2129] was built as a replacement for these. The trend however was towards wagons of higher capacity and relatively few were built. The example shown is to the standard design for this class, although a few have spoked wheels. [*Liskeard, Cornwall 1970*]

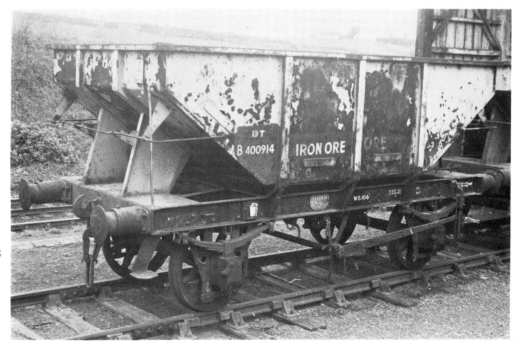

25½-TON NON-FITTED SAND HOPPER WAGON [STANDARD DESIGN] B439336 [Built 1957; BR Shildon; Lot 3001] is an example of the second BR design of ore hopper wagon to appear. They have a rather more substantial appearance than the 24-ton design and are of welded construction. They are also used for iron ore traffic, the major proportion being allocated for this. A small number have roller bearings. Later stock was fitted with vacuum-brakes but many of these are allocated to special traffics such as limestone. [*Queenborough, Kent 1972*]

30-TON VACUUM-FITTED HOPPER WAGON [SPECIAL TYPE] This wagon, B438238 [Built 1956; BR Shildon; Lot 2733], has been converted from a standard 25½-ton ore hopper. Most of the modifications are visible but beneath the tarpaulin is a supporting framework. The use of two vacuum-cylinders is noteworthy. The load although not specifically indicated usually consists of chemicals which would be affected by rain. The stock is restricted to the Widnes area and is fairly numerous in that locality. [*Widnes, Lancs. 1971*]

33-TON VACUUM-FITTED IRON ORE HOPPER WAGON [SPECIAL TYPE] This stock is very restricted in its area of use, being only found in the Glasgow area for conveying imported iron ore. There is some similarity between this type and the 25½-ton ore hopper but the sharply-sloping ends are distinctive. B445343 [Built 1957; BR Shildon; Lot 2962] is typical. Present-day working of these vehicles involves block-trains, usually composed of twelve wagons, hauled by a pair of Class 20 diesels. [*Glasgow (General Terminus), 1972*]

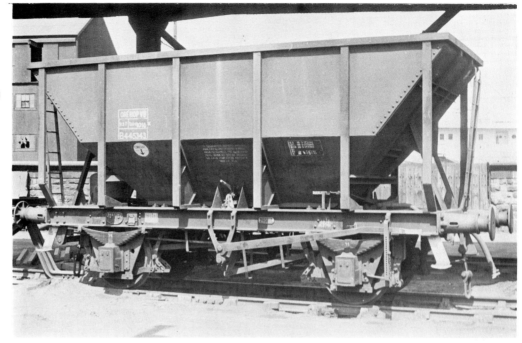

25-TON VACUUM-FITTED ANHYDRITE HOPPER WAGON [SPECIAL TYPE] For a class of wagon which is very restricted both in load and area of use, this stock achieved fame towards the end of steam on BR, being marshalled into the now well-known 'Long Meg' trains which ran between the Settle & Carlisle line and Widnes. All the wagons are similar to B747115 [Built 1954; BR Shildon; Lot 2597]. A point of interest to the intending modeller is the yellow tarpaulin. [*Newbiggin, 1970*]

26-TON IRON ORE TIPPLER WAGON [STANDARD DESIGN] Various of the iron and steel works opened since nationalisation have wagon tippler facilities and, to cater for these, the class of wagon shown here was introduced. It resembles the 16-ton mineral design but lacks any form of door. The bulk of the production was of non-fitted stock but some of the later batches have vacuum-brakes, as illustrated by B388639 [Built 1961; BR Derby; Lot 3363]. The use of roller bearings is not confined to vacuum-fitted stock. Recently repainted wagons have lost the large legend in the centre of the body and some are now allocated to stone traffic. [*Stainby, Lincolnshire 1969*]

13-TON SAND TIPPLER WAGON [STANDARD DESIGN] The sand tippler wagon is virtually a half-size iron-ore tippler. All were built as vacuum-fitted and B746591 [Built 1951; BR Swindon; Lot 2267] is a typical example of the earliest batches, later wagons having modernised brakegear and hydraulic buffers. The use of spoked wheels is no longer common. These vehicles were concentrated at the sand-pits in Staffordshire and Cheshire, although some were also allocated to Leighton Buzzard. Route instructions on the central panel are in yellow on black, with a yellow border. [*Congleton, Cheshire 1968*]

13-TON CHINA CLAY OPEN WAGON [WITH END DOOR: STANDARD DESIGN] To replace and supplement a rather decrepit fleet of china clay open wagons inherited from the GWR, BR introduced a vacuum-fitted design with an end door to facilitate unloading into the hold of a ship via a tipping platform. This stock stays mainly within ex-GWR territory although some are seen in the Midlands and on Southern Region in Kent. B743243 [Built 1956, BR Swindon; Lot 2871] is typically white in appearance although painted bauxite. Loaded wagons are normally sheeted, often with a blue plastic tarpaulin. [*Par, Cornwall 1972*]

11-TON CONFLAT L WITH L-TYPE CONTAINERS [STANDARD DESIGN] Prior to nationalisation, most cement was carried in bags and loaded in ordinary vans. The first attempt at bulk transit was the L-type container which has top loading and bottom discharge. These containers are painted grey with white lettering on black panels. The carrying vehicle is a Conflat L, a vacuum-fitted wagon easily identified by holes in the floor and projections around these for positioning the containers. B733583 [Built 1954; BR Ashford; Lot 2671] has the hinged sides found on Lowfits, their use on a Conflat L being rather obscure. Other Conflat Ls lack any form of side or end. [*Paisley (Ferenslie), 1972*]

22-TON PRESSURE-DISCHARGE BULK POWDER WAGON [PRESFLO: SPECIAL TYPE] The L-type container, although an improvement on previous practice, was not a very efficient method of carrying bulk powder, the main drawback being the wasted wagon space. The Presflo represented a big advance, carrying 10 tons more for only a small increase in length. Most Presflos are allocated to cement traffic but others are used to carry salt, slate powder or alumina. B887811 [Built 1958; Gloucester C & W; Lot 3177] shows the basic features of the type but is actually a special wagon which is permitted to travel on European railways. [*Hoo Junction, Strood 1969*]

20-TON PRESSURE-DISCHARGE BULK POWDER WAGON [PRESTWIN] Experience with the Presflo design showed that the load had a tendency to collect in the square corners of the body, thus preventing a completely free discharge. To overcome this problem, the Prestwin class was introduced, based on continental practice and slightly longer than the Presflo. Their main use is now on sand traffic, many cement companies having purchased or hired their own fleet of cement-carrying wagons. B873748 [Built 1962; Central Wagon Co., Lot 3469] is typical of the type. [*Congleton, Cheshire 1968*]

24-TON BULK POWDER WAGON [COVHOP: STANDARD DESIGN] This class of wagon was introduced in early BR days and has become the standard stock for conveying powdered chemicals. Many of them, as illustrated by B886017 [Built 1952; BR Derby; Lot 2375], are allocated for 'Soda Ash' traffic and may also be seen at oil refineries. Both non-fitted and vacuum-fitted wagons have been produced, some of the latter having roller-bearings. [*Grain, Kent 1968*]

13-TON BULK POWDER OPEN WAGON [SPECIAL TYPE] This type, built in small numbers, is a conversion of the first pattern of all-steel open wagon built by BR to an LNER design. The major difference concerns the security-bars which prevent the accidental lowering of the doors. They also have tarpaulin bars, a feature also found on ordinary all-steel open wagons of this type. B489177 [Built 1953; Birmingham R.C.W.; Lot 2361] is allocated to sand traffic but others are used for soda ash. [*Sittingbourne, Kent 1969*]

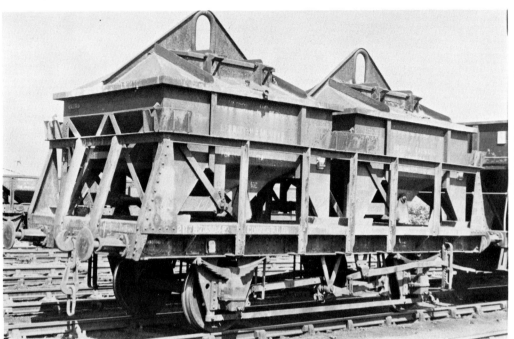

21-TON CONFLAT LD WAGON WITH LD-TYPE CONTAINERS [STANDARD DESIGN]

LD-type containers are especially for carrying dolomite, a form of limestone. The large oval hole on top of the container is the location point for crane hooks or slings when removing the containers. The Conflat LD is a standard 15′ wheelbase non-fitted wagon with special framework for the containers. The livery of both Conflat and containers is grey with white lettering on black panels. This example is B738042 [Built 1953; BR Swindon; Lot 2633].
[*Morpeth, County Durham 1970*]

13-TON CONFLAT A WITH LF-TYPE CONTAINER [STANDARD DESIGN]

LF19881B is a special container for carrying flour in bulk. Discharge is by suction. The livery is white with black lettering, the legend on the upper half of the container being 'BULK FLOUR CONTAINER'. B505256 [Built 1958; Pressed Steel; Lot 3153] is a late-pattern Conflat A which has been fitted with additional framework to position the LF-type container. [*Downham Market, Suffolk 1970*]

20-TON BULK GRAIN COVERED HOPPER WAGON [STANDARD DESIGN] Both the LMS and GWR produced steel-bodied covered hopper wagon types for bulk grain traffic and BR-built stock was developed from these. Early wagons have rivetted bodywork but the majority have the welded pattern illustrated by B885351 [Built 1956; Pressed Steel; Lot 2925]. Later examples are vacuum-fitted and roller-bearings are common on both non-fitted as well as vacuum-fitted stock. [*Kings Lynn, Norfolk 1970*]

21-TON BULK GRAIN OPEN HOPPER WAGON [SPECIAL TYPE] This stock is a modification of the standard BR design of 21-ton Coal Hopper and B419213 [Built 1955; BR Shildon; Lot 2854] is one of a small number built to this modified design. The most prominent feature is the tarpaulin bar, unusual for a BR design in being fixed. When carrying grain regularly this wagon would have been sheeted even when empty but, when recorded, was in fact carrying a mineral load which did not require a tarpaulin. [*Hoo Junction, Strood 1968*]

13-TON OPEN WAGON [MEDIUM-HEIGHT; DROPSIDE: STANDARD DESIGN] The LMS produced wooden-bodied vehicles of this type during the early 1930's. BR continued construction, the first vehicles built also having wooden bodywork. The bulk of BR construction however was the all-steel vehicle illustrated here by B458293 [Built 1951; BR Ashford; Lot 2235]. Two forms of brakegear were used, both of which were vacuum-fitted. One load recorded for this class is container traffic but all have now been transferred to the Engineers Department, where the drop-sides are a useful feature. [*Hoo Junction, Strood 1970*]

13-TON OPEN WAGON [HIGH-SIDED; ALL-WOODEN BODYWORK: TYPICAL DESIGN] All companies were producing high-sided open wagons with wooden bodywork in the years prior to nationalisation and BR produced further examples. B477285 [Built 1950; BR Ashford; Lot 2153] is a typical example. With the easing of the post-war steel shortage, BR turned to building open wagons with either steel ends or all-steel bodywork. All the wooden-bodied stock was vacuum-fitted and had modern buffers, a small number also having tarpaulin-bars. [*Edge Hill, Liverpool 1971*]

13-TON OPEN WAGON [HIGH-SIDED; WOODEN-BODIED WITH STEEL ENDS]

To combat the damage caused by shifting loads damaging the end-planks of open wagons, the LMS developed open wagons with corrugated steel ends. BR continued production of these and a high proportion of BR-built open wagons have this feature. As with the 13-ton Medium Dropside type, two forms of brakegear appeared and the type illustrated is the standard early pattern BR vacuum-brakegear. Hydraulic buffers, the 'Oleo' type, are seen here on B495051 [Built 1952; BR Ashford; Lot 2409]; tarpaulin bars are found on some examples. [*Hoo Junction, Strood 1968*]

13-TON OPEN WAGON [HIGH-SIDED; ALL-STEEL BODYWORK: LNER DESIGN]

For its 'Green Arrow' fast vacuum-fitted freight service, the LNER built open wagons with distinctive all-steel bodywork. BR produced large numbers of similar stock and B480215 [Built 1949; BR Shildon; Lot unknown] is one of the first batch, which were copied from the LNER design. Later batches had standard BR vacuum-fitted brakegear and, later still, vehicles with slightly modified bodywork appeared. The example shown has spoked wheels, uncommon in later years. [*Hoo Junction, Strood 1971*]

12-TON SHOCK-ABSORBING OPEN WAGON [HIGH-SIDED; WOODEN-BODIED] Shock-absorbing open wagons were produced by the SR and LMS during World War II and B721064 [Built 1950; BR Ashford; Lot 2155] is based on the SR type. These vehicles are noticeably taller than other shock-absorbing open wagons because the floor is the same width as the solebar. Many shock-absorbing vehicles, both open wagons and ventilated vans, are allocated to special traffics and this example is for use by Pilkington Brothers, the well-known glass manufacturers. [*Edge Hill, Liverpool 1971*]

12-TON SHOCK-ABSORBING OPEN WAGON [HIGH-SIDED; WOODEN-BODIED WITH STEEL ENDS] The LMS shock-absorbing open wagon design had the same corrugated steel end used on their ordinary open wagon and this design became the BR standard. B724180 [Built 1955; BR Derby; Lot 2766] is typical of the early BR-built stock. All early vehicles have the short buffers seen here and some have LMS-pattern brakegear. The type shown is, however, the more common. The route-instruction, which is in yellow on black with a yellow border, indicates that this wagon is allocated to the John Summers steelworks at Shotton, Flints. [*Strood, Kent 1968*]

12-TON SHOCK-ABSORBING OPEN WAGON [STANDARD DESIGN; LATE VARIANT]
Most BR-built stock dating from the early-1950's had design features which had been in use for many years. From the mid-1950's onwards improved buffer and brake-gear designs were introduced and B725521 [Built 1958; BR Derby; Lot 3082] has Oleo hydraulic buffers and eight-shoe vacuum-brakegear. This particular wagon, although retaining the white stripes and code-name of a shock-absorbing vehicle, has had the springs removed from below the doors. [*Pilning, Somerset 1972*]

13-TON SHOCK-ABSORBING OPEN WAGON [HIGH-SIDED; ALL STEEL BODYWORK] As stated earlier, only the LMS and SR produced shock-absorbing open wagons. The appearance in 1949 of a shock-absorbing version of the LNER 'Green Arrow' all-steel open seems to indicate that the LNER had also planned to introduce the type. The design retained the distinctive LNER-pattern vacuum-brakegear but the clean lines of the original design were marred by the crude method of joining the body to the floor. The only batch built, illustrated here by B720358 [Built 1949; BR Darlington; Lot 2033], comprised just over 500 vehicles. [*Hoo Junction, Kent 1970*]

12-TON VENTILATED VAN [VACUUM-FITTED; GWR PLYWOOD DESIGN] The final design of ventilated van constructed by the GWR before nationalisation had plywood bodywork and B753188 [Built 1949; BR Swindon; Lot 2079] is one of the small batch of BR-built vans which used the same design. The vacuum-pipe fitted to this van is a common feature of GWR-type vans. The sides and roof-profile of this design were incorporated into the BR standard ventilated van design. [*Whittlesea, Hunts. 1969*]

12-TON VENTILATED VAN [VACUUM-FITTED; SR PLYWOOD DESIGN] B752970 [Built 1949; BR Ashford; Lot 2063] is one of the small number of ventilated vans built by BR which followed the final variant of standard SR ventilated van design, which allied the distinctive Maunsell features of this with plywood bodywork. Two patterns of vacuum-brakegear are to be found on this stock. The less common type is illustrated here and follows SR practice. The majority have the type found on B753188, illustrated above. [*Aylesbury, Bucks. 1971*]

12-TON VENTILATED VAN [VACUUM-FITTED; LMS PLANKED DESIGN] The LMS were producing vans to a number of different designs just prior to nationalisation and BR built batches of vehicles to two of these. Both had typical LMS features, such as corrugated steel ends and sliding doors, but differed in the bodywork material used. The variety with planked bodywork is illustrated here by B751707 [Built 1949; BR Wolverton; Lot 2003]. The vacuum-brakegear used is a type commonly found on LMS-built stock but the tall vacuum-brakepipe is normally associated with LNER-built vans. [*Dinton, Wilts. 1969*]

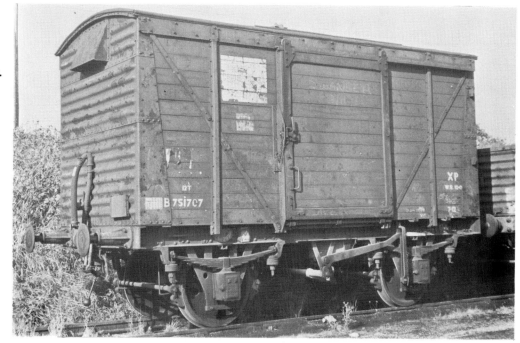

12-TON VENTILATED VAN [VACUUM-FITTED; LMS PLYWOOD DESIGN] Similar in all respects except for the material used to construct the sides and doors, B750086 [Built 1949, Wolverton, Lot 2001] represents the plywood-bodied LMS-pattern ventilated van. It is obvious from the angle of the photograph just where the inspiration for the corrugated steel ends, found on BR standard vans, came from but actually the LMS-pattern vans are taller and have a shallower roof curve. [*March, Cambs. 1969*]

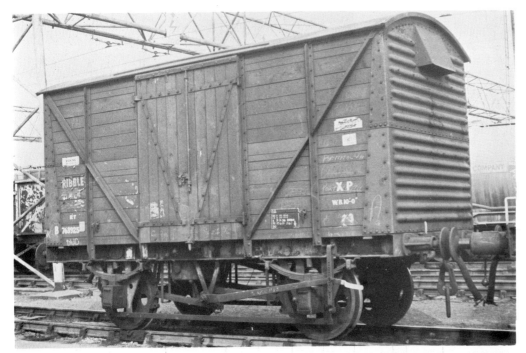

12-TON VENTILATED VAN [VACUUM-FITTED; STANDARD DESIGN, FIRST PATTERN] B763295 [Built 1955, BR Wolverton, Lot 2414] is a late example of the first pattern of BR standard ventilated van to be introduced. Large numbers, in excess of 10,000, were built to this particular design. A number of the early batches have tall vacuum-pipes more commonly found on GWR vans. In the mid-1950's, the second pattern of standard van appeared and these were identical except for the sides and doors, which were made of plywood. [*Hoo Junction, Strood 1971*]

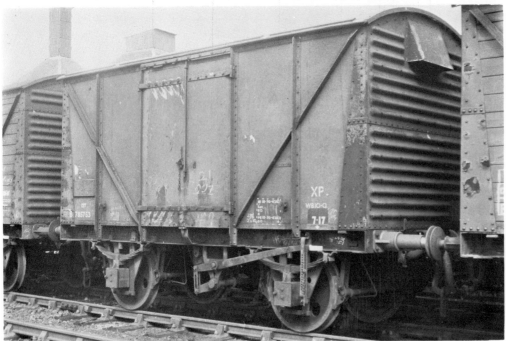

12-TON VENTILATED VAN [VACUUM-FITTED; STANDARD DESIGN, FOURTH PATTERN] As was referred to under shock-absorbing open wagons, eight-shoe vacuum-brakegear and hydraulic buffers were introduced on many BR designs. The first ventilated vans to have them appeared in 1957 and the third pattern of standard van combined these features with wooden sides and plywood doors. The fourth pattern reverted to all-plywood bodywork; B785753 [Built 1961; Pressed Steel; Lot 3398] is one of these. Some vans of the third and fourth patterns have Oleo hydraulic buffers instead of the type seen here. [*Coupar Angus, Perthshire 1971*]

12-TON VENTILATED VAN [VACUUM-FITTED; WITH WIDE-OPENING DOORS]

This class of vehicle can be regarded as one of the forerunners of the new van types recently introduced by BR. The doors are the significant feature of the design because they allow loading by fork-lift more easily than previous designs of ventilated van. All were built to the design illustrated by B784676 [Built 1962; BR Wolverton; Lot 3391] although some have been experimentally fitted with translucent plastic roofs and are lettered accordingly.
[*Trafford Park, Manchester 1968*]

12-TON SHOCK-ABSORBING VENTILATED VAN [VACUUM-FITTED; GWR DESIGN]

Although the LMS and SR built open wagons with shock-absorbing equipment, these vehicles were actually preceded into service by a GWR shock-absorbing ventilated van design. BR-built vehicles were similar to the GWR pattern but plywood replaced the planked bodywork B850185 [Built 1950; BR Ashford; Lot 2158] is representative of the type, although the use of spoked wheels can be considered uncommon. Approximately 500 of this type appeared. [*Hoo Junction, Strood 1970*]

12-TON SHOCK-ABSORBING VENTILATED VAN [VACUUM-FITTED; LNER DESIGN] The LNER seems to have been rather slower to adopt new ideas in wagon-design than were the other companies and had not introduced production batches of shock-absorbing stock before nationalisation. BR built, to LNER design, both open wagons and ventilated vans with such equipment. Less than 100 of the van type were produced and B850074 [Built 1949, BR Darlington, Lot 2045] is a typical example. [*Avonmouth, Somerset 1971*]

12-TON SHOCK-ABSORBING VENTILATED VAN [VACUUM-FITTED; LMS DESIGN] B850000 [Built 1950, BR Wolverton, Lot 2014] is one of the very few shock-absorbing vans built to this typical LMS-pattern design. It is possible that neither this type nor the LNE-inspired design featured above were perpetuated because the strain placed on the right-hand end of the springing when the door was in the open position. [*Hoo Junction, Strood 1971*]

12-TON SHOCK-ABSORBING VAN [STANDARD DESIGN; HYBRID VARIANT] The BR standard design of shock-absorbing ventilated van has the same features as the ordinary ventilated van. Both planked and plywood-bodied examples have appeared, four-shoe vacuum-brakegear being standard, although a small number appeared with LNER-pattern eight-shoe vacuum-brakegear. Most of these have planked bodywork with plywood doors but B850658 [Built 1951; BR Darlington; Lot 2202] is rarer in having the reverse. [*Dinton, Wiltshire 1971*]

12-TON SHOCK-ABSORBING VENTILATED VAN [STANDARD DESIGN; THIRD PATTERN] After the rather unusual construction of the above vans, design followed the pattern of the ordinary van design and innovations appeared during the late 1950's. However, only plywood-bodied stock appeared with hydraulic buffers and eight-shoe vacuum-brakegear. B854770 [Built 1959, BR Darlington, Lot 3117] has Oleo-pattern buffers but many have the slimmer design. This particular van has had its shock-absorbing equipment removed. [*Hoo Junction, Strood 1969*]

12-TON VENTILATED PALLET VAN [VACUUM-FITTED; WITH OFFSET DOORS]

During the late 1950's, a need existed for vans which could be loaded with wooden pallets. Some GWR ventilated vans were converted initially but later on BR built a large batch of new vehicles. Of these, B781910 [Built 1960; BR Wolverton; Lot 3310] was one of the first to appear. A serious derailment led to doubts on the riding characteristics of the type and they were soon withdrawn from service. A few have been modified and remain in general traffic but most have been condemned; many can now be seen in goods yards used as stores. [*Hartlepools, Co. Durham 1969*]

12-TON VENTILATED PALLET VAN [VACUUM-FITTED; WITH SLIDING DOORS]

Produced concurrently with the Wolverton-designed vehicles, the stock illustrated here by B782360 [Built 1960; BR Derby; Lot 3318] was far more successful. This design has an 11' wheelbase in contrast to the Wolverton-built batch [10' wheelbase] and this may be why their riding characteristics are somewhat better. Longer vehicles have also appeared and the development of the new air-braked vans can be traced from this type. [*Earlestown, Lancs. 1971*]

12-TON SHOCK-ABSORBING VENTILATED PALLET VAN VACUUM-FITTED] Small numbers of this class of van were built, with considerable variation, between the various batches. The standard corrugated steel end, absent in the ordinary pallet van designs, was fitted on all batches but the door seen on B855131 [Built 1961; BR Wolverton; Lot 3311] is only one of the types used. Another can be seen on the van to the right. [*Chorley, Lancs. 1972*]

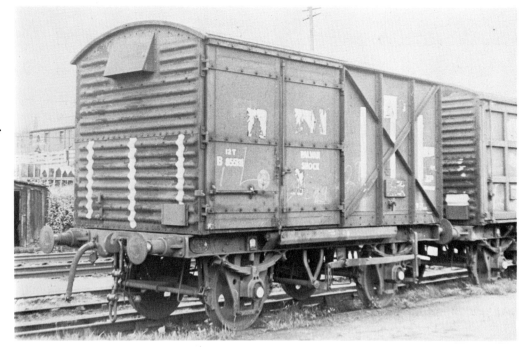

24-TON LONG-WHEELBASE AIR-BRAKED PALLET VAN [EXPERIMENTAL] Although BR's first attempt at catering for palletised traffic was not a conspicuous success, the idea was not abandoned and was developed eventually into the new long-wheelbase van, COV AB. Certain experimental designs were tried out before ordering the production batch of COV AB's, of which B787479 [Built 1966; BR Ashford; Lot 3590] is one. This particular experiment is unusual in not having fixed sides. The tarpaulins which stretch from roof to solebar are light grey. [*Newton Abbot, Devon 1971*]

11-TON CONFLAT A, WITH A-TYPE FURNITURE CONTAINER Container traffic was extensively developed in the 1930's by the railway companies in an attempt to combat the threat from road haulage. The A-type is the smaller standard size for covered containers and A2775B is a BR-built furniture container, derived from LMS practice, with one end door. The carrying vehicle is B704739 [Built 1957; BR Swindon; Lot 2971] and is a standard first-pattern Conflat A, developed from GWR practice. This method of loading is normal for A-type containers but it is possible for two such containers to to be loaded on to one Conflat A. [*Morpeth, Co. Durham 1970*]

13-TON CONFLAT A, WITH B-TYPE FURNITURE CONTAINER The major standard size of container, the B-size, was used by all companies. Containers with one end-door only are classified B-type and B9811B is a BR-built example with a steel end similar to A2775B, illustrated above. B503829 [Built 1958; Pressed Steel; Lot 3153] is a standard second-pattern Conflat A with Oleo hydraulic buffers and eight-shoe vacuum-brakegear. With the introduction of the Freightliner service, small-sized containers have become redundant and many, such as this example, have been condemned. [*Hoo Junction, Strood 1971*]

13-TON CONFLAT A, WITH BK-TYPE FURNITURE CONTAINER Externally similar to the B-type container, the BK-type has additional internal straps and hooks for load-security, BK9186B being a plywood-bodied BR-built example. As with the B-type, only one end-door is fitted. Note the large number of patches applied to damage on the visible side. B504203 [Built 1958; Pressed Steel; Lot 3152] is a standard second-pattern Conflat A [*Hoo Junction, Strood 1969*]

MODIFIED 13-TON CONFLAT A, MODIFIED BD-TYPE FURNITURE CONTAINER. By far the most numerous type of container, the BD-type has one end door and one door per side. Most were built to the LMS design which has planked bodywork and a steel end. BD47140B has, however, been modified for use on the pioneer LM Region 'Speedfreight' service and is in a livery of light grey with a yellow band and black lettering. The Conflat A, B737779 [Built 1959; BR Ashford; Lot 3107], has also been modified. [*Forres, Inverness-shire 1971*]

MODIFIED 13-TON CONFLAT A, WITH SPEEDFREIGHT CONTAINER A4338B is an example of the standard LM Region 'Speedfreight' service container. The livery is natural metal with a yellow band and black lettering. This service was the direct forerunner of the 'Freightliner' system, the similarity between these containers and the later Freightliner containers being most marked. The modified Conflat A is B737743 [Built 1959; BR Ashford; Lot 3107]. [*Lostock Hall, Lancs. 1970*]

12-TON VACUUM-FITTED CATTLE WAGON [SR DESIGN; MODIFIED VEHICLE] Many of the cattle wagons acquired from the companies at BR's formation were non-fitted and, to replace these, batches of vacuum-fitted stock were built immediately. Most of this early-period stock was built to an existing LMS design. However, a small number were built to an SR design and B891364 is one of these. This particular wagon has been modified for use as a tunnel inspection vehicle in the Standedge Tunnel. [*Marsden, Yorks. 1968*]

8-TON VACUUM-FITTED CATTLE WAGON [STANDARD DESIGN] BR used the GWR cattle wagon design as its pattern when producing a standard type for this class. B894228 [Built 1954; BR Swindon; Lot 2495] is part of a batch of smaller vehicles. Other batches were rated 11 tons but, although slightly taller, were alike in other respects. Few examples of this class remain in service, most cattle traffic being moved by road today. Those that do remain are concentrated on the seaports for Ireland and the Scottish islands. [*Chester 1968*]

10-TON VACUUM-FITTED VENTILATED MEAT VAN For moving animal carcasses which do not require insulation, BR built 200 standard Ventilated Meat vans. The design is a modification of the standard ventilated van, the most prominent features being the end ventilators and side louvres. None now remain on their original duties and some, when demoted to ordinary traffic, have been modified, one recently seen having all but the top end ventilators removed. B870074 [Built 1952; BR Wolverton; Lot 2320] is un-modified except for livery. [*Earlestown, Lancs. 1970*]

BM-TYPE VENTILATED MEAT CONTAINER BD11293B is an example of the container equivalent of the Ventilated Meat van. Construction was similar to the van in that the planked sides have louvres. These latter are also found on the ends, the fixed-end positioning being the same as the door-end. When made redundant from their original traffic, this type, by virtue of the fact that it has one end door and one door per side, was re-classified BD—as this example has been lettered. [*Hoo Junction, Strood 1971*]

10-TON VACUUM-FITTED INSULATED MEAT VAN Most meat traffic is conveyed in insulated stock and such vehicles are much more common than the ventilated types. B872187 [Built 1953, BR Wolverton, Lot 2321] is an example of the BR-standard design and, as will be seen, is a modification of the standard van. This particular example, original livery, is still in use as a perishable van, although carrying fish in this instance. Most have been relegated to ordinary traffic and have been repainted. [*Grimsby, Lincs. 1970*]

FM-TYPE INSULATED FISH & MEAT CONTAINER Similar in overall shape to the BM-type container, the FM-type was originally introduced by the LMS. BR also built large numbers and FM60388B is a typical plywood-bodied BR-built example. Until the late 1960's they were in regular use but they are now being phased out of service due to the loss of this traffic to road haulage. B505461 [Built 1958; Pressed Steel; Lot 3153] is a standard second-pattern Conflat A [Hoo Junction, Strood 1969]

12-TON INSULATED FISH VAN [PASSENGER; LNER PATTERN] In its final years, the LNER introduced a class of long-wheelbase insulated fish van which ran to passenger train timings on the East Coast main line. BR built further batches of this type, the initial design being similar to the LNER-built stock but having roller-bearings instead of oil-axleboxes. With the loss of much fish traffic to road haulage, many of these vans have become redundant and some have been converted to Special Parcels Vans, as illustrated by E87160 [Built 1954, BR Darlington; Lot 30125]. The livery is Rail blue with white lettering. [Stranraer 1972]

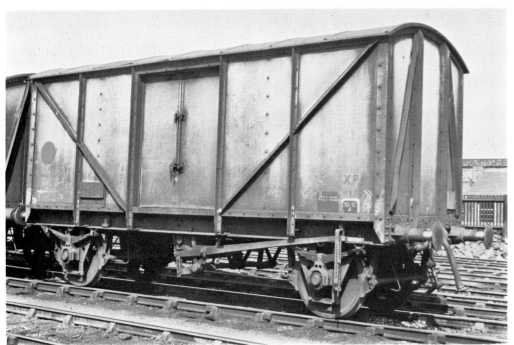

12-TON INSULATED FISH VAN [PASSENGER; STANDARD PATTERN]
Further batches of insulated fish vans were built to a modified design which, whilst retaining the same basic dimensions, had modernised vacuum-brakegear and GWR-style roof profile and ends. E87713 [Built 1959, BR Darlington; Lot 30384] has retained its original livery complete with the 'Blue Spot' which gives this class its nickname. Some of this later batch have also been relegated to Special Parcels Vans, in a similar manner to E87160, illustrated previously. [*Grimsby, Lincs. 1969*]

11-TON CONFLAT A WITH AFU-TYPE INSULATED CONTAINER
Insulated A-type containers, coded AF, were built in some numbers and were fairly common on perishable trains. AFU16327B is a BR-built container fitted with special equipment for carrying frozen foodstuffs. Many large firms, such as Birds Eye, had containers marked with their own brand-names and symbols. This example is, however, in the standard BR livery for insulated stock. The carrying vehicle, B703760, is a standard first-pattern Conflat A, uncommonly fitted with spoked wheels. [*Lostock Hall, Lancs. 1968*]

12-TON VENTILATED FRUIT VAN [VACUUM-FITTED; GWR DESIGN] B875272 [Built 1949, BR Swindon; Lot 2084] is one of a small batch of fruit vans built to a modification of the GWR Goods Fruit A class. They differ by having plywood ends and doors whereas the original was all-planked. A noteworthy retention from the original design is the small step beneath the left-hand door. Many of this type, and indeed the other fruit van types, are used for ordinary traffic and may be seen anywhere in Britain. One van of this class has been noted with the slats replaced by plywood sheets. [*Hoo Junction, Strood 1969*]

12-TON VENTILATED FRUIT VAN [VACUUM-FITTED; LMS DESIGN] The LMS built an experimental fruit van in the 1940's which was similar to their all-plywood ventilated van design. BR, however, built a batch of ventilated fruit vans which were modifications of the LMS ventilated van design with plywood sides and steel ends. The only basic difference between this class and the ordinary van is the addition of scoop ventilators just above solebar level. This modification was seen again in the BR standard Ventilated Fruit van. [*Whittlesford, Cambs. 1970*]

12-TON VENTILATED FRUIT VAN [VACUUM-FITTED; LNER DESIGN] Ventilated fruit vans were built in some numbers by the LNER and BR also built a batch of vans which followed the final all-plywood design used by the LNER. B754589 [Built 1950; BR Darlington; Lot 2134] is one of these. They were exact copies of the original design, the only noted variations being in the length of vacuum-pipe on some vans. The number series is unusual for this class and may indicate that they were either ordered as, or converted from, ordinary ventilated vans. [*Hoo Junction, Strood 1968*]

12-TON VENTILATED FRUIT VAN [VACUUM-FITTED; STANDARD DESIGN] After producing small batches of vans to pre-nationalisation designs, BR produced a variation of the standard ventilated van design for fruit traffic. The only difference was the addition of small scoop-type ventilators just above solebar level. B875772 [Built 1957; BR Darlington; Lot 3009] is typical of its class and —as can be established from the posters—has just carried two separate loads of cement and one of parcels. [*South Lynn, Norfolk 1970*]

8-TON INSULATED BANANA VAN [VACUUM-FITTED; LNER PATTERN] The conveyance of bananas has been part of the railway scene for many years and still continues today. BR built large numbers of new vans, the first standard design being derived from LNER practice. This design was in production long enough to receive such modern features as hydraulic buffers and eight-shoe vacuum-brakegear; B881548 [Built 1958; BR Darlington; Lot 3119] is one of the later vans. A noteworthy point for the modeller is the painting in black of the tare weight on the yellow spot which identifies the class. [*Hoo Junction, Strood 1968*]

12-TON INSULATED BANANA VAN [VACUUM-FITTED; STANDARD DESIGN] The late 1950's saw the introduction of a new-style banana van, with plywood bodywork and corrugated steel ends. B881804 [Built 1959; BR Wolverton; Lot 3209] is typical of the class and no design variations have been observed. Individual vehicles have occasionally been seen in goods yards not associated with banana traffic and, due to the introduction of an air-braked design making them redundant, they may possibly be used in ordinary traffic.
[*Lostock Hall, Lancs. 1969*]

12-TON GUNPOWDER VAN [STANDARD DESIGN] BR built batches of gunpowder vans which were almost identical to GWR-built stock, thus continuing a precedent set in pre-grouping days when most companies copied the GWR 'Iron Mink' for their gunpowder van design. BR built batches of both non-fitted and vacuum-fitted stock, construction continuing into the late 1950's. B887135 [Built 1959; BR Swindon; Lot 3237] is one of the final batch and has such modern features as hydraulic buffers and eight-shoe vacuum-brakegear. Gunpowder vans can be observed in all parts of the country. [*Mistley, Essex 1970*]

22-TON PLATE WAGON [STANDARD DESIGN; NON-FITTED] The normal load for vehicles of this class is sheet metal in flat form which is less than normal wagon-width. Plate wagons are common in many areas but, as with all steel-carrying stock, more so in industrial and steel-producing districts. B931173 [Built 1951; BR Shildon; Lot 2199] is typical of non-fitted stock although its spoked wheels are not common. Some non-fitted Plates have received roller-bearings. A number of Double Bolster wagons have been converted to Plate configuration and these can be identified by the B92XXXX series number. [*Cadder, Mid-Lothian 1972*]

22-TON PLATE WAGON [STANDARD DESIGN; VACUUM-FITTED] The first vacuum-fitted Plate wagons appeared in the mid 1950's and had oil-axleboxes and spring-buffers. Later stock has hydraulic buffers and two patterns of enclosed axlebox. The more common type is the roller bearing but some have the BR-patented 'Hybox', as illustrated by B935050 [Built 1960; BR Shildon; Lot 3223]. The load on this particular wagon is not so unusual as might be thought; the drop-sides of the class allow vehicles to be driven on from a loading-dock.
[*Wrexham, Denbighshire 1969*]

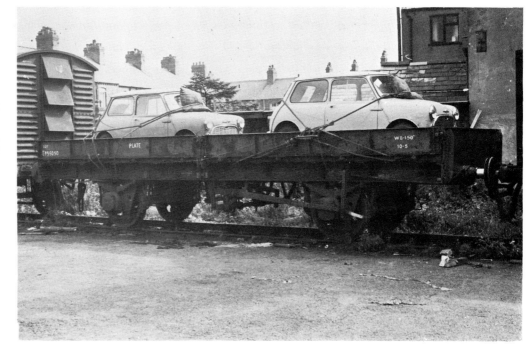

42-TON BOGIE PLATE WAGON [NON-FITTED; STANDARD DESIGN] Longer sheets of steel which are inside the normal wagon width are carried by bogie plate wagons. The LMS had vehicles of this type and BR-built a batch to a similar design, B947050 [Built 1950; BR Shildon; Lot 2133] being one of these. Later non-fitted stock had the GWR-pattern bogie found on many BR built bogie wagons. Some Bogie Bolster D wagons have been converted to Boplate E and these can be identified by having rigid sides and numbers in the B942XXX series.
[*Liverpool Docks 1971*]

42-TON BOGIE PLATE WAGON [VACUUM-FITTED; STANDARD DESIGN] The vacuum-fitted version of the Boplate E appeared in the late 1950's and B948386 [Built 1960; BR Ashford; Lot 3240] is typical of the majority of wagons in this class. Some have the Gloucester-designed bogie which has appeared on a number of late-period BR bogie designs. All vacuum-fitted Boplate E wagons have brake-wheels, as seen on this example, to operate the hand-brakes. [*Hereford 1972*]

21-TON TRESTLE PLATE WAGON [NON-FITTED; 4-WHEELED TYPE] Trestle wagons are used to convey steel sheets which are wider than normal wagon width, carried on a framework at an angle of 45°. Both 4-wheeled and bogie vehicles have been built. The 4-wheeled stock resembles a Plate wagon but, in addition to the wooden framework, one side is removed. The LNER had similar wagons. All BR-built stock has been converted, either from Plate wagons or, as is B920016 [Built 1949; BR Shildon; Lot 2020], Double Bolster wagons. [*Stainforth, Yorks. 1969*]

50-TON BOGIE TRESTLE WELL WAGON [VACUUM-FITTED]

In addition to various examples of Bogie Trestle wagons which were built by BR, some bogie Well wagons were also fitted with trestle framework for carrying extra-large plates. These are classified Trestrol, both non-fitted and vacuum-fitted examples having appeared. B901739 [Built 1960; BR Derby; Lot 3309] is a typical example of the latter and retains the basic Weltrol configuration as well as having the trestle framework. [*Liverpool Docks 1971*]

13-TON SHORT-WHEELBASE PIPE WAGON [WITH DROP-SIDES]

BR produced a small number of 10′-wheelbase open wagons for pipe traffic. These had drop-sides and both non-fitted and vacuum-fitted versions were produced, B483706 [Built 1949; BR Ashford; Lot 2061] being one of the latter. The provision of hydraulic buffers on a design of this vintage is unusual. These wagons are no longer used for their original traffic and have been transformed to the Engineers Department. [*Hoo Junction, Strood 1971*]

12-TON PIPE WAGON [LNER DESIGN] The LNER produced 12′-wheelbase drop-side Pipe wagons mainly for use at the Stanton & Stavely Ironworks and BR made this its standard design for the traffic. B740539 [Built 1949; BR Darlington; Lot 2047] is one of the first batch, which were copies of the LNER vehicles. Later, modifications were made to the brakegear. Some vehicles of this class are not numbered in the standard series, the odd batch being in the B484XXX series. [*Rochester, Kent 1969*]

22-TON TUBE WAGON [VACUUM-FITTED; LNER DESIGN] Tube traffic consists of long steel pipes of varying diameters and both the LMS and LNER built special vehicles for it. The LNER produced all-planked stock in both fixed-side and drop-side forms. BR produced a batch of LNER-style drop-side vehicles with a 19′-wheelbase and B730406 [Built 1949; BR Darlington; Lot 2049] is one of these. This particular wagon has unusually high ends but is otherwise a standard example. All were vacuum-fitted and some have BR 'HYBOX' axleboxes or roller-bearings. [*Corby, Northants. 1970*]

22-TON TUBE WAGON [NON-FITTED; GWR DESIGN] This type was classified 'OPEN C' in GWR days and was not regarded as being solely for tube traffic. BR classified all ex-GWR Open C's as Tubes and built further examples to the final 4-plank, 19' 6''-wheelbase design. All were non-fitted, B731078 [Built 1950; BR Swindon; Lot 2127] being typical; some received roller-bearings. [*Cliffe, Kent 1971*]

22-TON TUBE WAGON [NON-FITTED; STANDARD DESIGN] BR perpetuated the LMS tube wagon design as its own standard and large numbers were produced. In overall length, the LMS-derived design is longer than either the LNER and GWR types but the wheelbase (17' 6'') is shorter. The first batches comprised non-fitted stock and B731471 [Built 1953; BR Swindon; Lot 2328] is standard for its class. Some have received roller-bearings. [*Edge Hill, Liverpool 1971*]

22-TON TUBE WAGON [VACUUM-FITTED; STANDARD DESIGN]

Vacuum-fitted variants of the standard tube wagon were first put into service in the late 1950's. Initially ordinary oil-axleboxes were fitted but later both BR 'HYBOX' axleboxes and roller-bearings were used; B733163 [Built 1959; BR Darlington; Lot 3226] has the latter. Hydraulic buffers with extra-large heads are standard. Many have wooden slats fixed transversely on the floor to raise the load and these are classified 'TUBE(BATTEN)'. [*Hoo Junction, Strood 1969*]

13-TON SINGLE BOLSTER WAGON [NON-FITTED; STANDARD DESIGN]

Most companies have built Single Bolster wagons and the BR standard vehicle was based on an all-steel design produced by the LNER. B910420 [Built 1949; BR Shildon; Lot 2034] is part of the first batch to be built. These early vehicles had a 9'-wheelbase but later ones had 10'-wheelbase and some were vacuum-fitted. This type is now redundant and, after brief use by the Engineers Department, the survivors are used on internal traffic at various large yards. [*Newport Docks 1972*]

22-TON DOUBLE BOLSTER WAGON [STANDARD DESIGN] This class is an extension of the Single Bolster idea and both the LMS and LNER built examples. The BR standard design is somewhat similar to LMS practice and can be considered a modification of the Plate wagon design. Many have been converted to either Plate wagon or Trestle wagon configuration. The survivors have been transferred to the Engineers Department where they are used in long-welded rail trains, B920109 (Built 1949; BR Wolverton; Lot 2020) being so used. [*Hoo Junction, Strood 1970*]

26-TON TWIN-BOLSTER WAGON SET [LOWFIT CONVERSION] Certain pre-grouping companies produced sets of permanently-coupled paired Single Bolster wagons. BR continued this idea and redundant stock of three basic types has been utilised. This example is of two Lowfit wagons, B451709 [Built 1952; BR Shildon; Lot 2340] and B453020 [Built 1959; BR Shildon; Lot 2998], which differ in the brakegear used on the individual wagons. Other conversions comprise pairs of vacuum-fitted Single Bolsters or pairs of converted Conflat A's. All conversions have the intermediate buffers and coupling hooks removed and replaced by a permanent coupling. [*Scunthorpe, Lincs. 1970*]

30-TON BOGIE BOLSTER WAGON [BBC; NON-FITTED]
The GWR had many Bogie Bolster wagons of 30-ton capacity and the standard BR design was similar in many respects to the GWR-built stock. Early BR-built wagons have diamond-frame bogies but the majority of the non-fitted ones have the plateback bogie-pattern normally associated with GWR bogie freight stock. B943134 [Built 1952; no other details known] is one of the later wagons but is unusual in having spoked wheels. [*Hartlepools Docks 1970*]

30-TON BOGIE BOLSTER WAGON [BBC; VACUUM-FITTED]
Considerable numbers of non-fitted BBC's were built and most are still in service. However, many of the steel trains in the 1970's consist of vacuum-fitted stock. The first vacuum-fitted BBC's were introduced in the late 1950's and production continued until the middle 1960's. The first batches have a modified form of plateback bogie, fitted with roller-bearings. Later batches have the Gloucester-patent bogie which is similar to that found on American stock. B924449 [building details not known] is one of the latter type and illustrates the fact that the bolsters are moveable, including the two end ones. [*Hoo Junction, Strood 1968*]

42-TON BOGIE BOLSTER WAGON [BBD; NON-FITTED] Whereas the 30-ton BBC was GWR-inspired, the 42-ton BBD followed LNER practice. Apart from the obvious difference in length, the main difference between the BBC and BBD is the ability, on the former, to move the bolsters. Thus the positioning of the two end bolsters on the BR-built BBD above the bogie centres, instead of at the end as found on an LNER-built BBD, hides the true parentage. The buffers and system of brake levers are purely LNER-inspired. As with BR-built BBC's, both diamond-frame and plateback bogie types were used, B941197 [Built 1950; Tees-side Eng.; Lot 2211] having the former. [*St. Neots, Hunts. 1970*]

42-TON BOGIE BOLSTER WAGON [BBD; VACUUM-FITTED] This class followed the pattern set by vacuum-fitted BBC's and two bogie types were used. The earliest pattern is seen here on B927605 [Built 1960; BR Lancing; Lot 3246] and later vehicles had the Gloucester pattern. A noteworthy point is the provision of handwheels to operate the hand-brakes. Note also the unusual form of bolster used at the nearest end of the wagon. The load is intended for use in the erection of re-inforced concrete structures and can be seen in any part of the country. [*Queenborough, Kent 1974*]

50-TON BOGIE BOLSTER WAGON [BORAIL] Borails are intended for conveying extra-long loads such as rails or, as seen here, concrete beams. BR inherited a variety of types from the companies and produced further new vehicles to a number of different designs. Many of these can be considered lengthened variants of Bogie Bolster wagon designs but the type illustrated here by B946229 [Built 1961; BR Derby; Lot 3334] has very distinctive bodywork. The bogies are unusually long for a British wagon and are only found on one other design, the BR Engineers Department 'Salmon' rail-carrying wagon. [*Hoo Junction, Strood 1969*]

13-TON SHORT-WHEELBASE VEHICLE-CARRYING OPEN WAGON [LOWFIT; LNER DESIGN] This class of wagon is generally used for conveying road vehicles or agricultural implements of suitable length. Most companies had similar vehicles in pre-nationalisation days but they lacked the drop-side and end facility which allows a road vehicle to be driven on from a loading dock. The LMS-built stock in this class had fixed wooden sides. The first batch of BR-built Lowfits were similar but, in addition to LMS van-type vacuum-brakegear, had hinged sides and ends. B450023 [Built 1950, BR Wolverton, Lot 2107] is loaded with a typical farm implement. [*Carmarthen 1971*]

13-TON SHORT-WHEELBASE VEHICLE-CARRYING OPEN WAGON [LOWFIT; STANDARD DESIGN] The standard BR-built Lowfit is an all-steel type with LNER characteristics. Early batches, as illustrated by B450782 [Built 1951; BR Shildon; Lot 2194], have LNER-pattern vacuum-brakegear. Subsequent batches had standard four-shoe pattern brakegear and eventually hydraulic buffers and eight-shoe brakegear. In addition to their vehicle-carrying duties, Lowfit wagons are also used as runners to shunting engines or as barriers for wagons with overhanging loads. [*Strood, Kent 1968*]

8-TON BOGIE VEHICLE-CARRYING OPEN WAGON [BOCAR P; EARLY CONVERSION] The Lowfit class is suitable for conveying individual road vehicles but the needs of the motor-manufacturing industry demand stock with higher capacity. When BR standard coaches began to appear in large numbers, the policy adopted decreed that pre-nationalisation stock be withdrawn even though it was not time-expired. As a result many such coaches were converted to car-carrying wagons. Early conversions, such as B889101 [Built 1925; LMS [Derby] and converted 1957; BR Wolverton; Lot 3090], were classified Bocar P. Most have been called Carflat A; coaches from all four companies, as well as some BR Mk. 1 coaches, have been converted. [*Radstock, Somerset 1971*]

25-TON VEHICLE-CARRYING WELL WAGON [LOWMAC WE; GWR DESIGN] Lowmac wagons are four-wheeled well wagons which have drive-on, drive-off facilities. All companies had wagons of this class and BR built further examples of each design. B905096 [Built 1957, BR Swindon, Lot 2975] is based on GWR-practice and even has the distinctive axlebox pattern associated with Swindon. The use of a well wagon is essential when the overall-height of the road vehicle, when loaded on to a flat wagon, is more than the railway loading gauge. The army vehicle shown is a typical load.
[*Carmarthen 1971*]

25-TON VEHICLE-CARRYING WELL WAGON [LOWMAC EU; MODIFIED LNER DESIGN] B904662 [Built 1955; Derbyshire C. & W; Lot 2714] is based on an LNER design but having additional modifications for Continental work—notably airbrakes, extra coupling-chains and special buffers. The lettering and 'anchor' symbol indicate that it complies with international regulations and can work on Continental railway systems. Comparing this with the Lowmac WE illustrated above, it will be seen that there are many design features which differ. The Lowmac class is possibly the least standardised of the BR fleet.
[*Strood, Kent 1970*]

38-TON VEHICLE-CARRYING BOGIE WELL WAGON [RECTANK EB]

This class is derived from a type of wagon produced by the War Department during World War I to carry the first army tanks, hence the name. A more usual load for the BR-built class is the military road-resurfacer seen on B909063 [Built 1960, BR Swindon, Lot 3299]. They were a late addition to BR's wagon fleet and some have been allocated to each region. A number of Eastern Region wagons have been recorded carrying sheeted loads of timber. [*Strood, Kent 1970*]

20-TON VEHICLE-CARRYING TROLLEY WAGON [FLATROL SB; MODIFIED DESIGN]

Because heavy track-laying vehicles such as bulldozers and cranes cannot pass over the wagon axles without exceeding the latter's permitted load, the sloping ends and drive-on, drive-off facility of the Lowmac is absent in the Flatrol. Most BR-built vehicles of the Flatrol class are non-fitted but B900030 [Built 1956, BR Lancing, Lot 2927] is modified for Continental working and therefore has airbrakes. The load in this case is a wagon which has been involved in a derailment and is being kept for examination by Board of Trade inspectors. [*Hoo Junction, Strood 1971*]

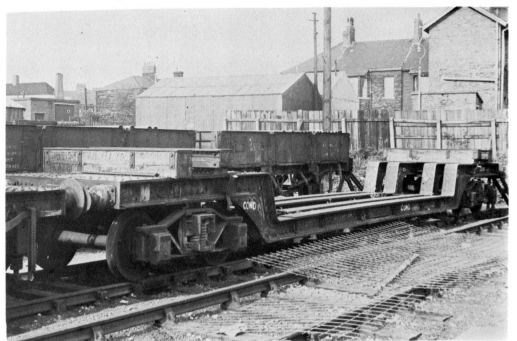

50-TON VEHICLE-CARRYING BOGIE TROLLEY WAGON [FLATROL EY] This larger type of Flatrol, illustrated by B900506 [Built 1953; BR Shildon; Lot 2475], is intended for cranes with overhanging frames or jibs. The timber baulks are removable and are stamped with the wagon number to make sure they return to the wagon. Most Flatrols are allocated to the Engineers Department, having DB-prefixed numbers, but both this vehicle and the Flatrol SB illustrated previously are Traffic Department stock. Some Flatrols have the code-name 'Loriot', perpetuating the GWR name for this class. [*Heaton Junction, Newcastle 1969*]

12-TON NON-FITTED GLASS-CARRYING WAGON [GLASS MD; LMS DESIGN] The carriage of crated plate-glass, not being a major source of traffic, is confined to journeys to and from Pilkington Bros. glassworks at St. Helens. The LMS, being the local company, built specialist stock for this traffic. BR produced further examples to the same design, including B902008 [Built 1953; BR Swindon, Lot 2491]. The GWR also had glass-carrying stock and these resembled a Lowmac wagon with upright frames. BR also built a small batch of vacuum-fitted vehicles to this design. [*St. Helens, Lancs. 1970*]

16-TON VACUUM-FITTED PALLET BRICK-CARRYING WAGON In the late 1950's a new traffic, consisting of bricks loaded on to wooden pallets, was contracted in Scotland and, to cater for it, BR converted a number of redundant Medfit open wagons. Most of the conversions are similar to B462184 [Built 1957; BR Ashford; Lot 2724] but some wagons retained the original Medfit ends. Although two capacities were built, 13 tons and 16 tons, they were externally similar. Most Palbricks are now themselves redundant and have been re-converted for further use on other duties. Two examples of new duties are 'Freightliner' Match wagon and Coil P, a steel-carrying type. [*Wrexham, Flintshire 1968*]

50-TON BOGIE TROLLEY WAGON [FOR CARRYING NUCLEAR WASTE; FLATROL MJ] The removal of nuclear waste in special lead containers from atomic power-stations, although new, is an increasing BR traffic. Initially it was carried by modified LNER-built Lowmac's but BR later built specially-designed stock of which B900527 [Built 1960; BR Swindon; Lot 3300] is typical. Some variation exists but the basic configuration of a long-bodied wagon with small central load-carrying area is retained. This traffic has kept open a number of lines which would otherwise have been closed; the gantry seen in this view is always a prominent feature of the loading point. [*Fairlie, Ayrshire 1972*]

15-TON VACUUM-FITTED TIMBER-CARRYING WAGON [STANDARD DESIGN] There are widely-separated localities having timber traffic on BR and each has its own wagon type. Firstly, there is the West Highland line which has considerable local traffic along it for pulp and paper mills. The wagons used were originally vacuum-fitted Plates, being first converted to Conflat wagons for London Midland Region's 'Condor' train and later into the current Timber P class, which has two separate load areas. The second standard type, illustrated by B455555 [Built 1962; BR Ashford; Lot 3465], can be found transporting imported timber, mainly from East Anglian ports or Portishead on the Bristol Channel. [*Boston, Lincs. 1969*]

22-TON VACUUM-FITTED ALE PALLET WAGON This class is a conversion of vacuum-fitted Tube wagons, intended to convey metal beer kegs on wooden pallets. Some variation occurs in the arrangement of the doors between the various individual conversions but B732693 [Built 1956; BR Darlington; Lot 2867] can be regarded as a typical example. Some have roller-bearings and others Hybox axleboxes. In addition to the conversions, some unconverted vacuum-fitted Tube wagons have been recorded with 'Bass-Charrington' plates, indicating that they have also been used for beer traffic. [*Chester 1968*]

VACUUM-FITTED DIESEL BRAKE TENDER [EARLY TYPE] Although rated 35 tons, these vehicles carry no pay-load and merely provide braking assistance to enable diesel locomotives to handle non-fitted or partially-fitted goods trains. Originally they were only to be in North-East England, usually propelled by the locomotive being brake-assisted. They can now be seen on all BR regions and current practice, at least on Southern Region, is to haul them. B964109 is an early example with square bodywork. [*Strood, Kent 1973*]

VACUUM-FITTED DIESEL BRAKE TENDER [STANDARD TYPE] Diesel brake tenders are the second example in this volume of the re-use of redundant assets, being re-built from surplus coaches. In this case, only the bogies are re-used in the conversion and the majority, including B964115 [Built 1964; BR York; Lot 3500] shown here, have ex-LNER Gresley-type bogies. Other examples may be seen with ex-LMS or BR Standard Mk. I coach bogies. The pattern of bodywork remains the same. [*Strood, Kent 1973*]

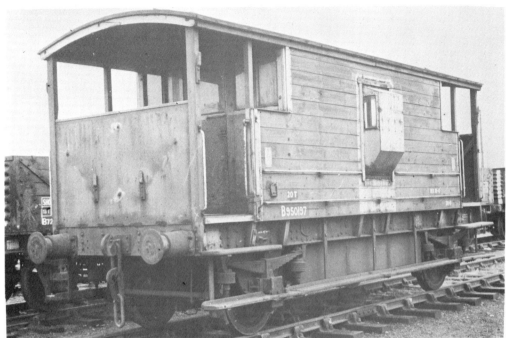

20-TON BRAKE VAN [NON-FITTED; LMS DESIGN] The type illustrated here was the final design of brake van built by the LMS and large numbers were constructed. BR produced further examples of both non-fitted and vacuum-fitted stock; B950197 [Built 1950; BR Derby; Lot 2026] is an example of the former. The rectangular extension beneath the solebar holds scrap metal to increase the weight and, therefore, the braking ability of the vehicle and is normally painted in the main body colour, in this case grey.
[*Hoo Junction, Strood 1968*]

20-TON BRAKE VAN [VACUUM-PIPED; LMS DESIGN] B950019 [Built 1950; BR Derby; Lot 2025] is a vacuum-piped example of the final LMS-standard brake van design, indicated by the white painted vacuum-pipe. On a brake van this allows the guard to slow the train by applying the vacuum-brakes on the wagons even though his own vehicle does not have vacuum-brakes. Some of this class, however, are vacuum-brake fitted and these have a red-painted vacuum-pipe, for identification.
[*Coatbridge, Lanarkshire 1972*]

20-TON BRAKE VAN [GWR DESIGN; NON-FITTED] The GWR standardised on its own distinctive brake-van design for most of its existence and BR built a small number of vehicles to the same design. B950608 [Built 1949; BR Swindon; Lot 2099] is one of the non-fitted vehicles built, a small number of vacuum-fitted vans also being constructed. The BR-built stock has always been rare, compared to those of GWR manufacture. When new, this particular van was allocated to Whitland, Cardiganshire.
[*Pen-y-Fford, Flintshire 1971*]

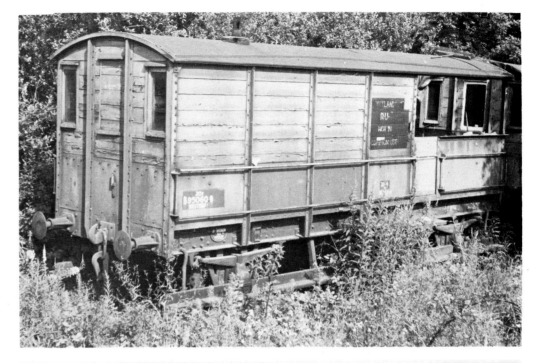

20-TON BRAKE VAN [STANDARD DESIGN; NON-FITTED] When it introduced the 'Green Arrow' fast fitted freight service, the LNER produced a variation of its standard brake van with a 16′ wheelbase and vacuum-brakes. With a number of detail differences, this was adopted as the BR standard design. Although the LNER did not produce any non-fitted examples, BR did and B950880 [Built 1950; BR Darlington; Lot 2137] is one of these. It has the various BR-inspired minor differences, such as concrete slabs on the end platforms, but retains the LNER practice of having two tail-lamp brackets. [*Avonmouth, Somerset 1972*]

20-TON BRAKE VAN [STANDARD DESIGN; VACUUM-FITTED] Vacuum-fitted examples of the BR standard brake van had a number of design variations between successive building runs. The first batch have tall buffer-beam vacuum-pipes whilst later batches have other pipes which project from solebar to body side. Later vehicles also have roller-bearings plus hydraulic buffers, including B955230 [Built 1962; BR Ashford; Lot 3394]. Many roller-bearing fitted brake-vans also have air-brakes. The lettering to the left of the look-out on this example is a legacy from GWR practice; a number of these vans have similar instructions. [*Hoo Junction, Strood 1969*]

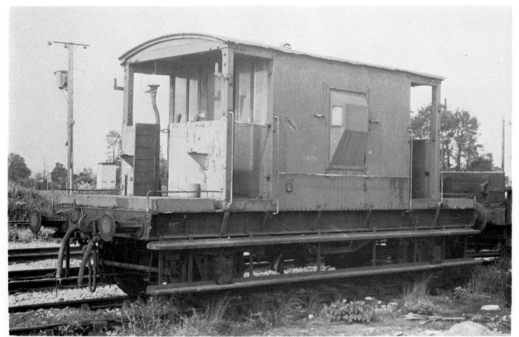

20-TON BRAKE VAN [MODIFIED STANDARD DESIGN] B951608 [Built 1952; BR Darlington; Lot 2349] is a specially modified brake van for use on trains which pass through the Severn Tunnel. Both ends have been similarly modified and the lower-half of the ends is painted yellow. Note the steam-heating pipe and lamp-bracket to the left of the buffer. [*Pilning, Somerset 1972*]